DISCARDED

Dragonflies
By Oxford Scientific Films

Photographs by George Bernard

G.P. Putnam's Sons   New York

First American Edition 1980
Text/Nature's Way copyright © 1980 by G. Whizzard Publications Ltd.
Photographs copyright © 1980 by Oxford Scientific Films Ltd.
All rights reserved.
Printed in Belgium by Proost, Turnhout

Library of Congress Cataloging in Publication Data
Oxford Scientific Films.
Dragonflies.
SUMMARY: Describes the physical characteristics and
habits of dragonflies,
which are among the oldest insects on earth.
1. Dragonflies — Juvenile literature.
[1. Dragonflies]
I. Bernard, George. II. Title.
QL520.09 1981   595.7'33   79-25942
ISBN 0-399-20731-7

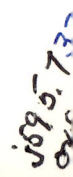

Dragonflies

Dragonflies (*Odonata*) are among the oldest insects on earth. Fossilized remains have been found showing that they existed 300 million years ago. More than 4,500 different species live in the world today. They vary in size and color, but they are all unmistakably members of the same enormous family. In the photographs that follow, we shall be looking at two common varieties of dragonfly (a *Libellulid* and an *Aeshnid*) and a close relative, the damselfly (*Zygoptera*).

Dragonflies and damselflies live near water, usually by stagnant ponds, marshes, slow-moving rivers or streams. They can be seen mostly during the late spring and summer. They have four large wings with a lace-like pattern of veins, long slender bodies, huge heads and prominent eyes. Most of them are brilliantly colored, with bodies that are red, blue, green, brown, yellow and so on. The color gets stronger as the insect gets older. Many dragonflies have a small, dark oblong mark on each wing tip. Other varieties have larger, more colorful markings. The color or pattern of the female is often different from that of the male, and usually more subdued.

The dragonfly is larger than the damselfly. Its wing span ranges from 5–10 cm (2–4 in.), with the back pair slightly broader and shorter than the front. As it flies, its wings make a rustling sound. Dragonflies move through the air at tremendous speeds, some species reaching 96 kmph (60 mph). They can fly for hours at a time and travel considerable distances, sometimes more than 32 km (20 miles). More often, though, they patrol a particular stretch of water looking for insects to eat, or they find a perch from which they can dart out to catch their prey. When resting, the dragonfly keeps its wings outspread at right angles to its body.

The more delicate damselfly is not such a strong flier or as fast, and it rarely moves far from its breeding place near water. Its four wings are the same size and, unlike the dragonfly, it rests with its wings raised above its body.

The dragonfly is mainly active during the day and relies a great deal on its sight. It has two enormous compound eyes which together almost cover the top of its head. Each eye is made up of about 30,000 facets or individual lenses. Each of these facets looks out in a slightly different direction from the others, collectively forming a kind of mosaic picture of the whole field of vision. The dragonfly can swivel its head on its thin neck to give it an all-round view of things.

A dragonfly's eyes are so sensitive to movement that it can spot and catch small insects, like mosquitoes, in full flight. The eyes of the damselfly are constructed in the same way, but they are smaller and positioned farther down the side of its head.

In contrast to its eyes, the dragonfly's antennae, which it uses for sensing touch and smell, are poorly developed and less important. They look like two small bristles on the top of the head. Its mouthparts, however, are very efficient. The mandibles, or jaws, have strong tooth-like projections for biting into the dragonfly's insect diet. Its Latin name *Odonata* means toothed.

Dragonflies have three pairs of legs, attached to the body just behind the mouth. These are used for seizing their prey in mid-air, and holding on to it while the jaws get to work. The legs are also used for clinging to plants and other objects, but dragonflies do not actually walk.

The body of the dragonfly, although generally long and narrow, does vary according to type. The *Libellulid*, for example, is noticeably broader and shorter than the *Aeshnid*. Damselflies are the smallest of all, often hardly thicker than a large darning needle. Most dragonflies are usually between 2.5–5 cm (1–2 in.) in length, but in some parts of the world there are considerably larger species.

The dragonfly's abdomen is divided into ten distinct segments, with males having a pair of claspers attached to the last one. On segments 2 and 3, the males have special reproductive organs. The mating process of dragonflies is unique among insects. Before joining up with the female, the male transfers sperm from the genital opening on segment 9 to the special organs on segments 2 and 3. He then seizes the female with his tail claspers, either by the head or the neck. Thus firmly gripped, the female curves her body around until her genital opening is touching the male's reproductive organs, and collects the sperm.

Mating usually takes place in the air. When it is over, the male releases the female to lay her eggs. Or, in the case of some species, the male continues to hold her during the egg-laying process. Damselflies and some dragonflies insert their eggs into the stems of water plants, or into mud. Other dragonflies simply scatter their eggs into the water, washing them off the tip of the abdomen as they skim over the surface. The number of eggs they lay varies from hundreds to thousands. Between two and five weeks later (longer with some species), the eggs hatch into nymphs. These have the basic body structure of the adult insect, but they are fatter and without wings. Their color is also quite different. Much the same variations in size exist, with the damselfly nymph the most slenderly built.

The nymphs remain underwater until they are ready to change into adult dragonflies and damselflies. They breathe by means of gills, through which they take in oxygen from the water. The gills of the dragonfly nymph are concealed within its rectum, but those of the damselfly can be clearly seen in the form of three long "tails."

Nymphs, like the adult insects, are carnivorous (flesh-eating). To help them catch their food they have developed a unique structure called the "mask." Their lower lip is greatly extended and hinged in the middle. On the end of it are two sharp hooks, like claws. When not in use the mask is folded back beneath the nymph's head and forelegs, hiding part of the face. (This explains the term "mask.") When the nymph sees food the mask is thrust forward with lightning speed, and the prey is caught on the hooks. The nymph draws the victim to its jaws, and eats it.

Nymphs are mostly a dull brown-green color, merging into the background. This helps to protect them from their enemies, such as large fish. It also makes it difficult for their own prey to see them until it is too late. Nymphs devour smaller insects and their larvae in great quantities. The larger dragonfly species will even tackle a tadpole and the occasional small fish, as well as smaller nymphs.

Apart from the development of the wing buds and the gradual enlargement of the eyes, there is little visible change during the nymphal life. In the case of damselflies, the life cycle is normally completed within one year. But dragonflies can take from one to five years, and possibly longer. During its life as a nymph the insect moults, shedding its skin, as many as ten or fifteen times.

When the nymph is ready to moult for the last time, it comes out of the water and climbs up a reed or some other perch above the surface. After a short rest, the skin splits and the head and chest emerge. Another rest follows before the abdomen is pulled clear of the old skin. The wings then expand and the abdomen stretches. In place of the dull squat nymph is a spectacularly colored dragonfly.

Many dragonflies and damselflies emerge early in the morning while it is still dark. Once they have been warmed by the sun, they fly away to enjoy their new, brief life. It lasts for just a month.

Some people believe that dragonflies can sting them. But this beautiful creature is harmless. Unless, of course, you happen to be one of the things it likes to eat.

Dragonflies and damselflies live near water, usually by ponds, marshes, slow-moving rivers or streams.

A *Libellulid*

There are many different species of dragonflies.

They vary in color and size.

An Aeshnid

Agrion splende[ns]

There are different kinds of damselflies too. They look much like dragonflies, but they have smaller wings and thinner bodies.

The dragonfly's enormous eyes almost meet on top of its head. It has excellent all-round vision and strong jaws for biting its food.

(*above*) These damselflies are about to mate. When they mate, the male damselfly grips the female's neck with his tail claspers.

The male holds onto the female as she enters the water to lay her eggs.

She places them inside the stem of a water plant.

Some dragonflies scatter their eggs on the surface of the water.

Two to five weeks later, the eggs hatch into larvae, or nymphs.

The nymph's large eyes are easy to see in this head of a nymph that has just hatched.

Many young nymphs are eaten by creatures such as this Great Water Beetle larva.

The dragonfly nymph has a short, fat body. It breathes air through gills in an opening at the rear of its body.

Damselfly nymphs are longer and narrower than dragonfly nymphs.

Their gills are outside the body in the form of three long "tails."

This dragonfly nymph chases a tadpole . . .

. . . and catches it by shooting out its extended lower lip, called a "mask."

Nymphs have big appetites and will even attack small fish like this stickleback.

The nymph's coloring helps it to hide from larger fish and other enemies.

A newly hatched dragonfly with the nymph shell underneath.

The dragonfly's wings before they expand.

When it is ready to change into a dragonfly, the nymph leaves the water and climbs onto a resting place.

The dragonfly warms itself in the sun before flying for the first time.

These damselflies are leaving their nymph shells together.

Their colors brighten as they dry.

This dragonfly has left its nymph shell behind.
As a nymph it may have lived for as long as five years.
As a dragonfly it will probably live for a month.

Books by Oxford Scientific Films

BEES AND HONEY
Photographs by David Thompson

THE BUTTERFLY CYCLE
Photographs by Dr. John Cooke

HOUSE MOUSE
Photographs by David Thompson

THE SPIDER'S WEB
Photographs by Dr. John Cooke

THE STICKLEBACK CYCLE
Photographs by David Thompson

THE CHICKEN AND THE EGG
Photographs by George Bernard & Peter Parks

COMMON FROG
Photographs by George Bernard

THE WILD RABBIT
Photographs by George Bernard

DRAGONFLIES
Photographs by George Bernard